Our Solar System

Illustrations: Janet Moneymaker
Design/Editing: Marjie Bassler

Our Solar System
ISBN 978-1-950415-44-1

Published by Gravitas Publications Inc.
Imprint: Real Science-4-Kids
www.gravitaspublications.com
www.realscience4kids.com

Earth is located inside a big **solar system.**

A solar system is made of a **star** and the group of **planets** and other objects that **orbit** the star.

Orbit is the name for the circular path a planet follows around a star.

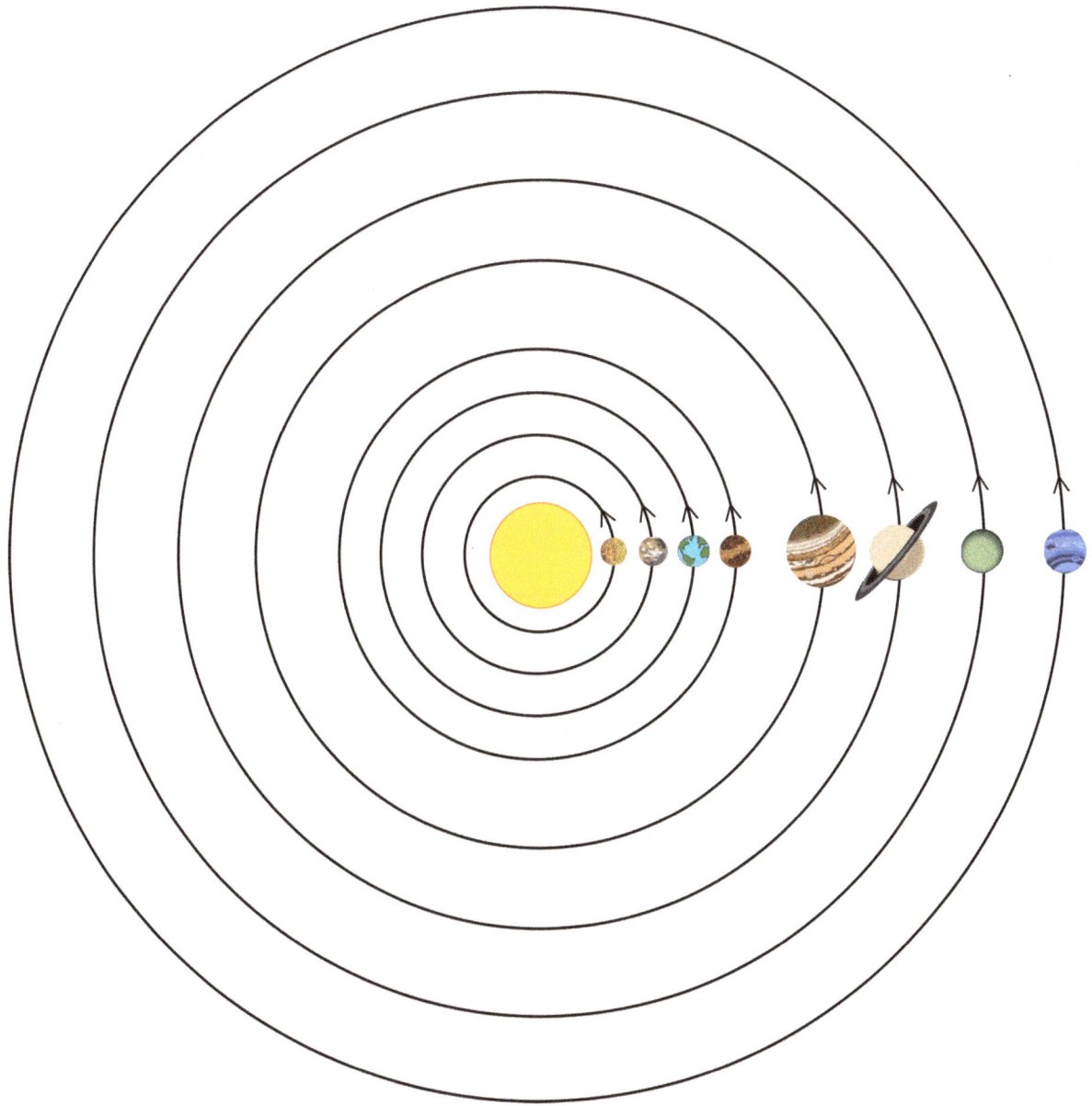

OUR SOLAR SYSTEM
The Sun and the planets

The Sun is the center of our solar system. The Sun is a **star**.

What is a star?

A star is an object in space that makes its own light and heat energy.

A photograph of the Sun

Our solar system has 8 planets, an **asteroid belt**, and lots of **moons** around the planets.

The eight planets in our solar system

SUN · MERCURY · VENUS · EARTH · MARS · JUPITER · SATURN · URANUS · NEPTUNE

The Asteroid Belt is here

The first four planets orbiting the
Sun are **Mercury, Venus, Earth,**
and **Mars.**

This group of planets is called the
inner solar system.

Look! Earth is
part of the inner
solar system.

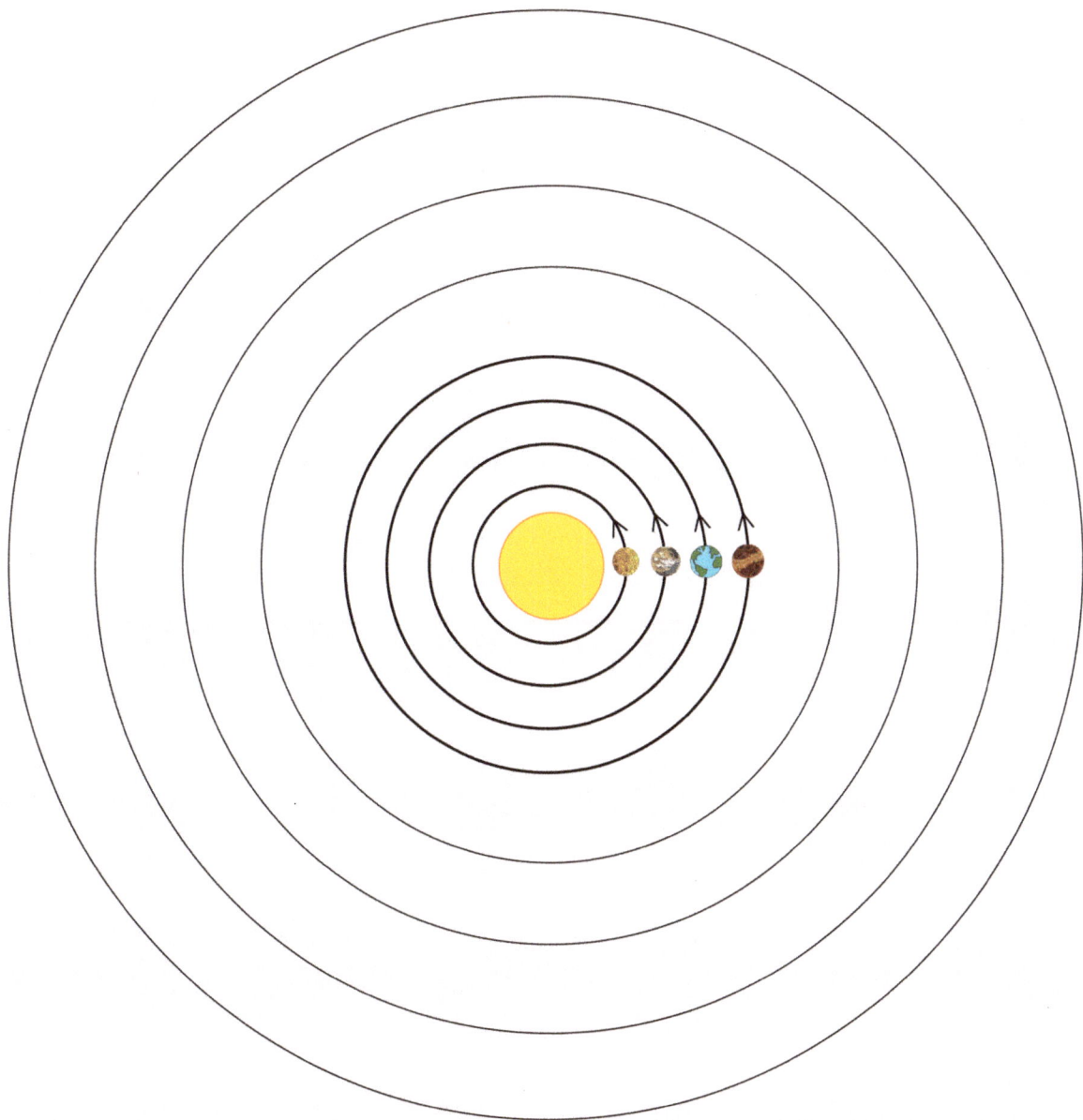

The next four planets orbiting the Sun are **Jupiter, Saturn, Uranus,** and **Neptune.**

This group of planets is called the **outer solar system.**

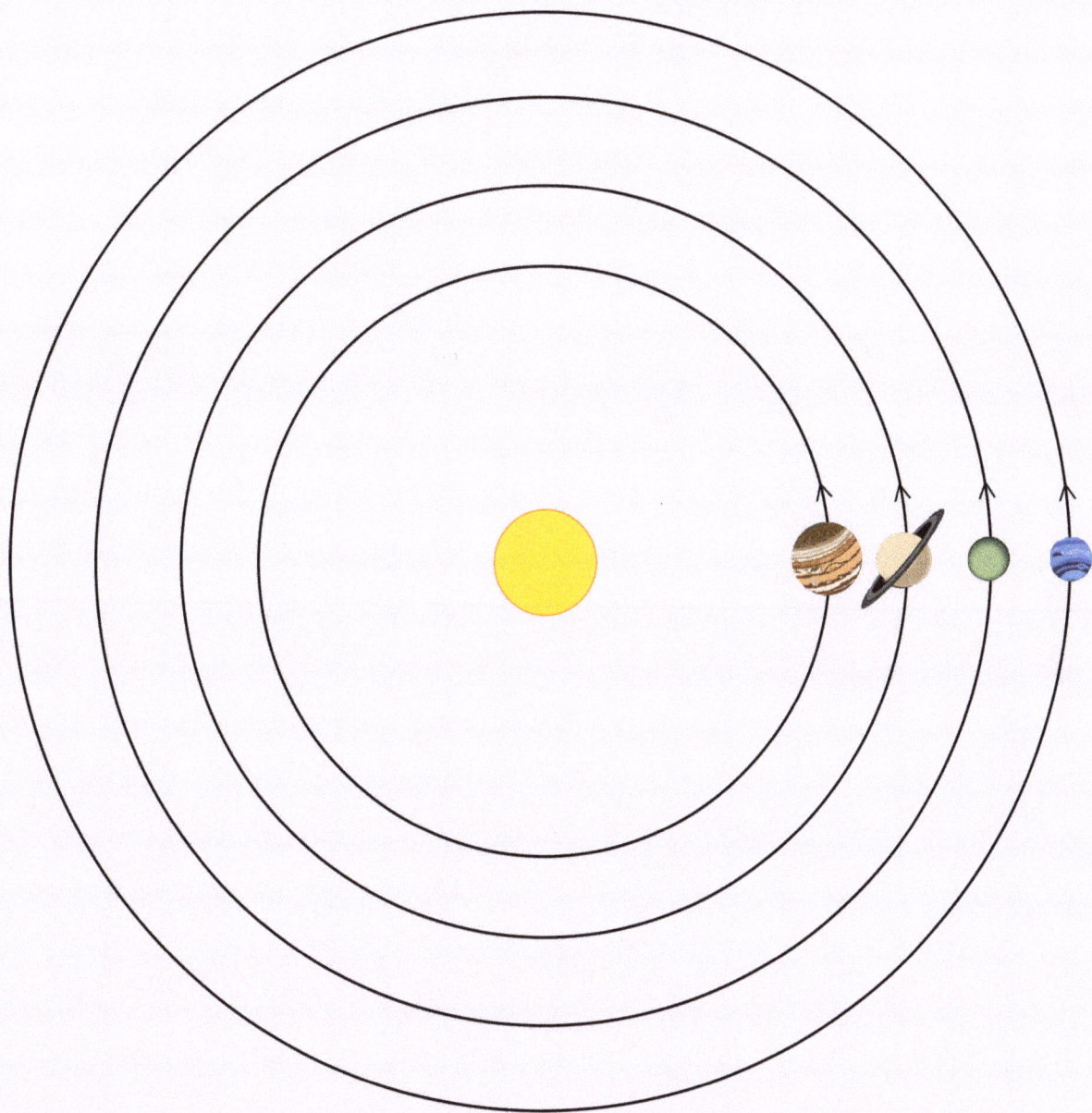

In between the inner solar system and the outer solar system is the **Asteroid Belt.**

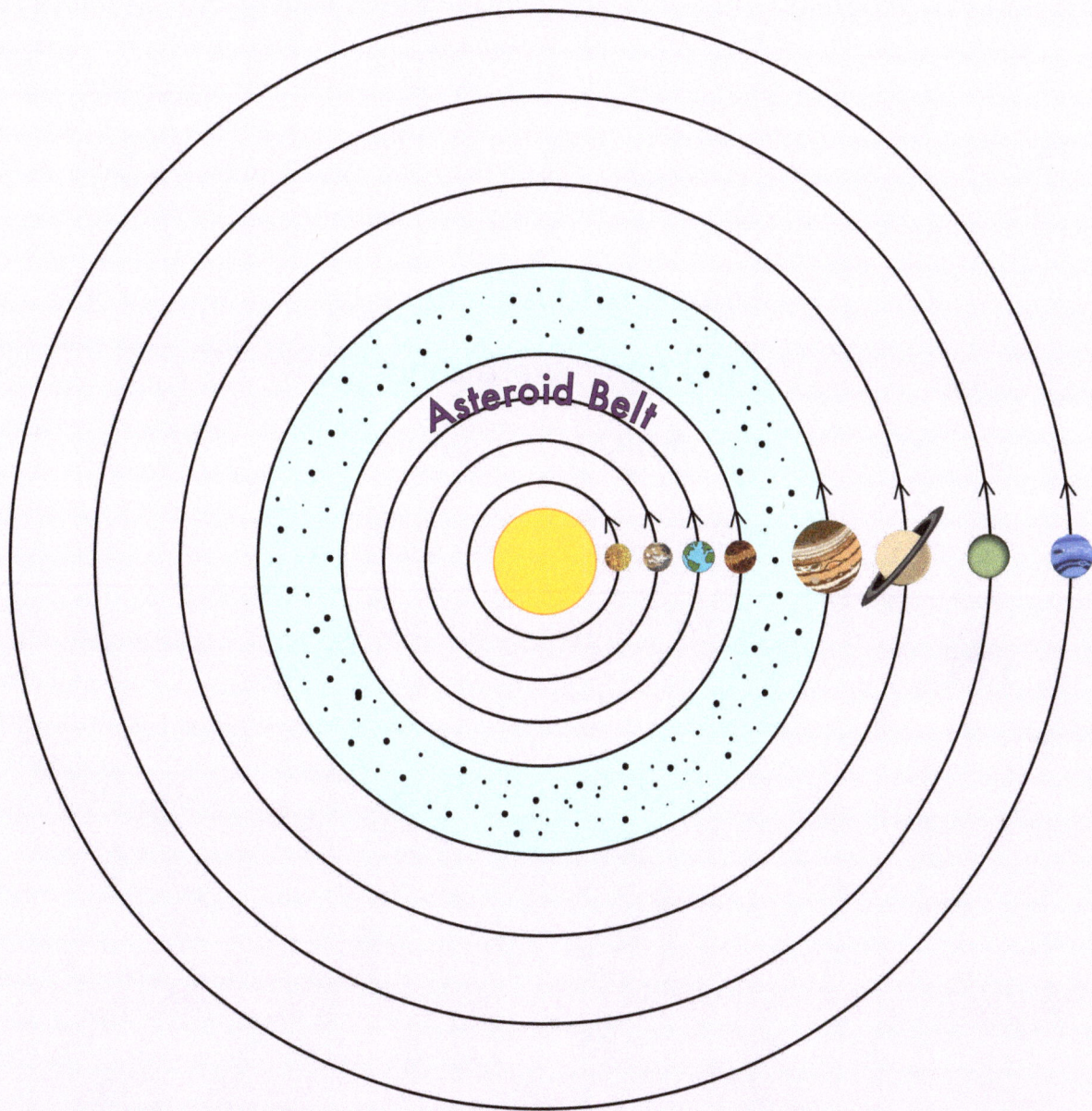

Asteroid Belt

An **asteroid** is a chunk of rocks and minerals that is too small to be a planet.

The Asteroid Belt is made of over a million asteroids that are orbiting the Sun.

Asteroids

Asteroid Vesta
Photo credit: NASA/JPL-Caltech/UCAL/
MPS/DLR/IDA

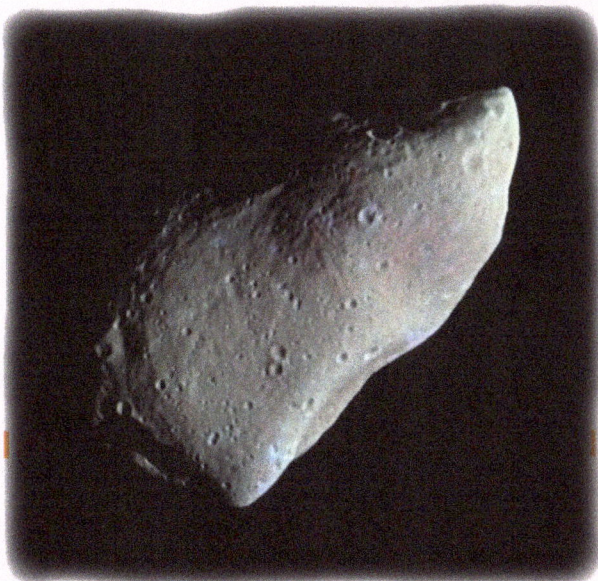

Asteroid Gaspra
Photo credit: NASA

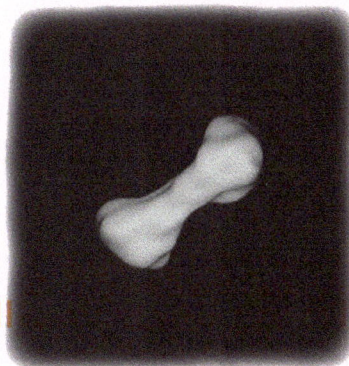

Asteroid Kleopatra
Photo credit: NASA

Not all asteroids are in the Asteroid Belt. Some asteroids come into Earth's air and burn up. When we see these, we call them **shooting stars** or **meteors.**

I love seeing shooting stars!

A meteor
Photo by Navicore, CC BY SA 3.0

Comets are also found in our solar system. A comet is made of ice, rocks, and dirt.

A comet is a dirty ice ball!

Hale-Bopp Comet

Planets with their moons, asteroids, and comets all circle our Sun to create the solar system.

I wonder how many other solar systems there are.

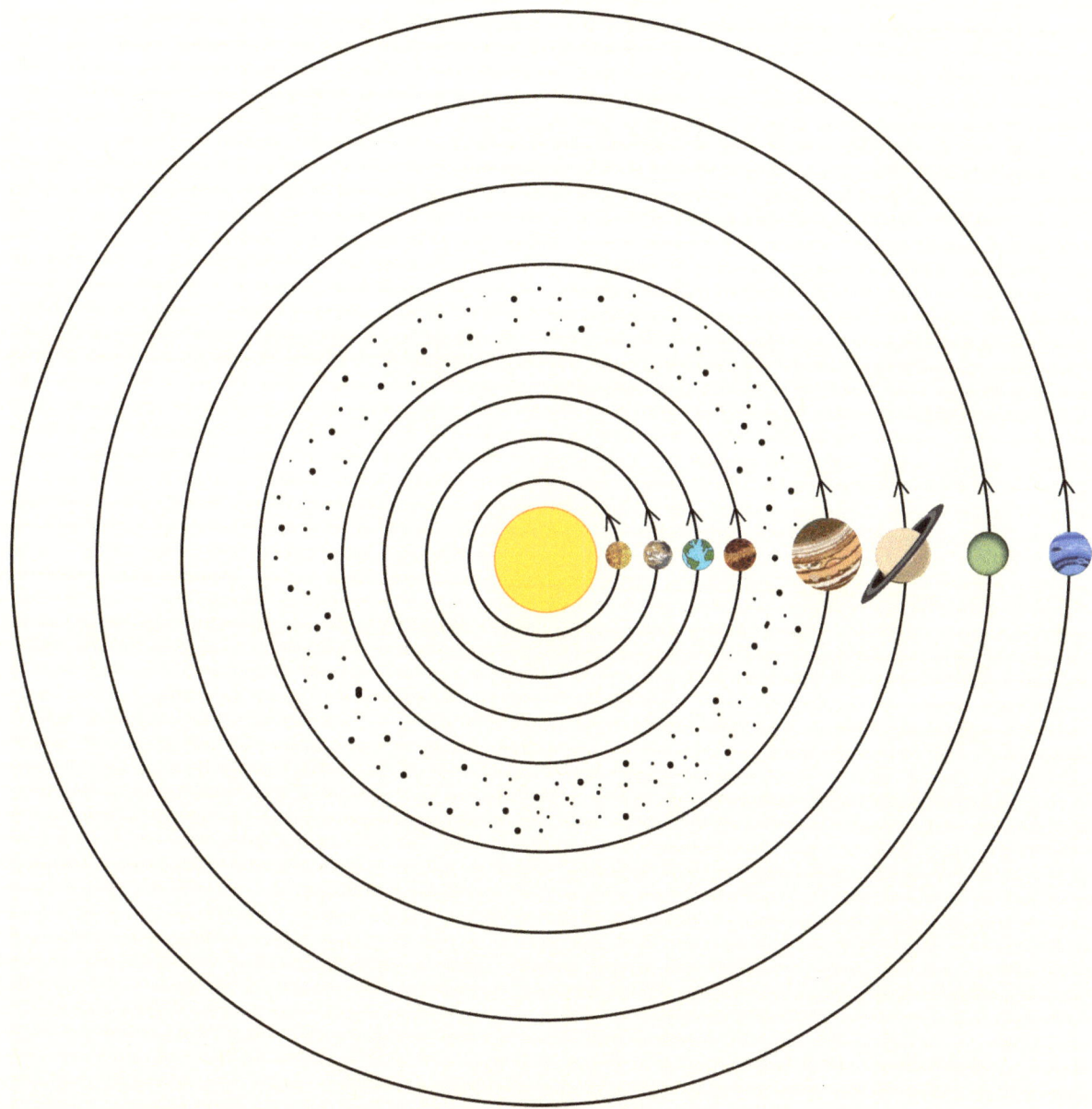

How to say science words

asteroid (AA-stuh-royd)

comet (KAH-muht)

energy (EN-uhr-jee)

Gaspra (GAS-pruh)

Hale-Bopp....(HAYL - BAHP)

Kleopatra (klee-uh-PA-truh)

meteor (MEE-tee-uhr)

orbit....(AWR-buht)

planet (PLA-nuht)

science (SIY-uhns)

solar (SOH-luhr)

space (SPAYSS)

star (STAHR)

system (SIS-tuhm)

Vesta (VE-stuh)

Planet Names

Earth (ERTH)

Jupiter (JOO-puh-tuhr)

Mars (MAHRZ)

Mercury (MUHR-kyuh-ree)

Neptune (NEP-toon)

Saturn (SA-tuhrn)

Uranus (YOOR-uh-nuhs)

Venus (VEE-nuhs)

What questions do you have about OUR SOLAR SYSTEM?

Learn More Real Science!

Complete science curricula from Real Science-4-Kids

Focus On Series

Unit study for elementary and middle school levels

Chemistry
Biology
Physics
Geology
Astronomy

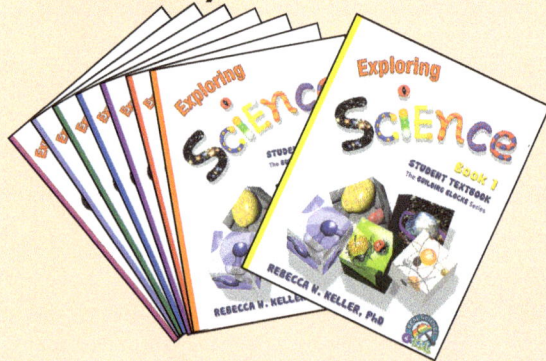

Exploring Science Series

Graded series for levels K–8. Each book contains 4 chapters of:

Chemistry
Biology
Physics
Geology
Astronomy